WHAT DOES IT LOOK LIKE?

# IT HAS PAWS!

BY KATHLEEN CONNORS

Gareth Stevens PUBLISHING

first concepts

Some animals
have paws.
Paws are feet
with claws.

Many pets have paws.

Dogs have paws.

Cats have paws.

Rabbits have paws.

Hamsters have paws.

Wild animals have paws too.

Tigers have paws.

Pandas have paws.

Foxes have paws.

Mice have paws.

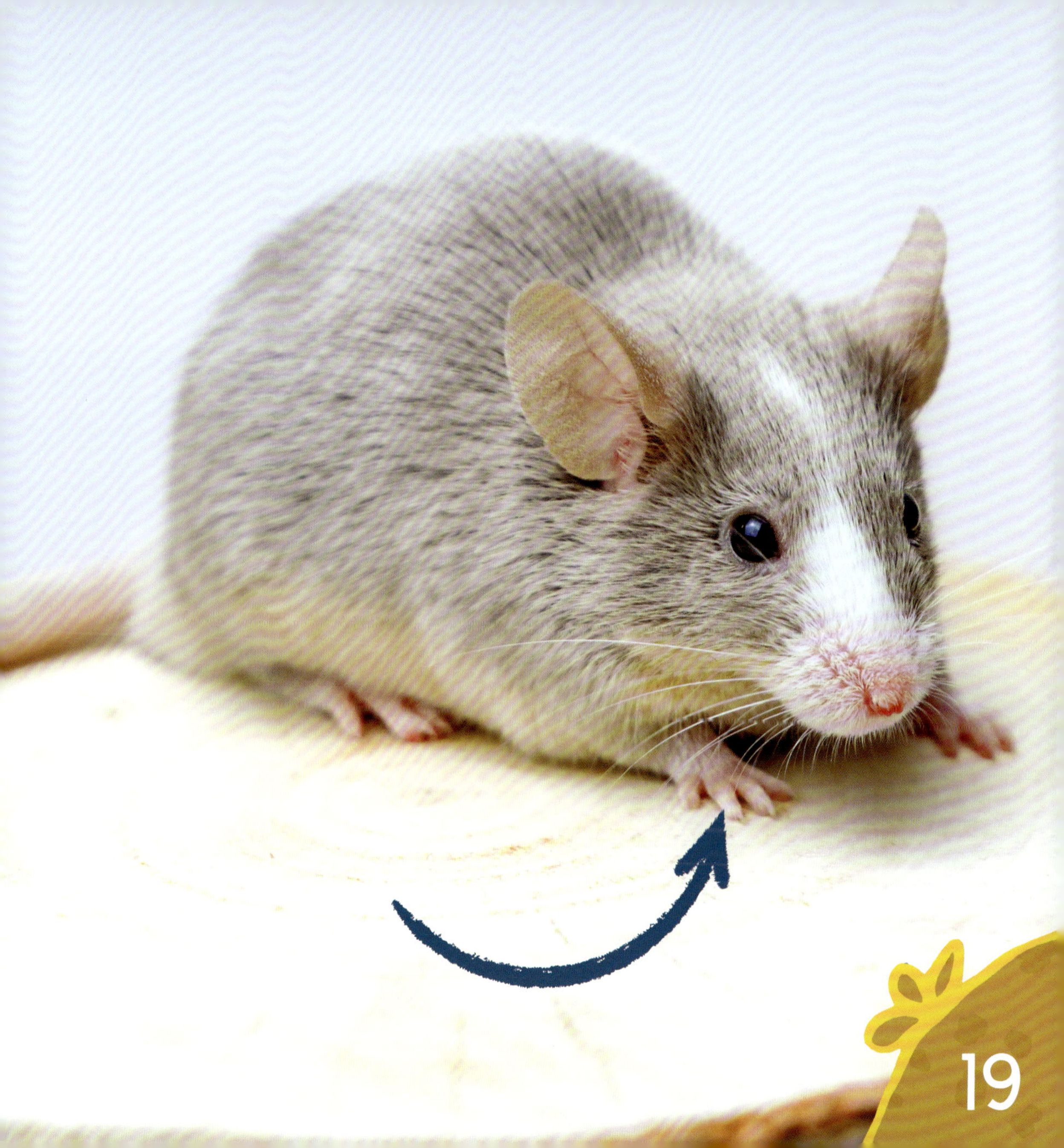

Raccoons have paws.

Which animal
has paws?

**Please visit our website, www.garethstevens.com. For a free color catalog of all our high-quality books, call toll free 1-800-542-2595 or fax 1-877-542-2596.**

**Cataloging-in-Publication Data**

Names: Connors, Kathleen.
Title: It has paws! / Kathleen Connors.
Description: Buffalo, NY : Gareth Stevens Publishing, 2026. | Series: What does it look like?
Identifiers: ISBN 9781482470901 (pbk.) | ISBN 9781482470918 (library bound) | ISBN 9781482470925 (ebook)
Subjects: LCSH: Foot–Juvenile literature. | Claws–Juvenile literature.
Classification: LCC QL950.7 C747 2026 | DDC 591.47'9–dc23

Published in 2026 by
**Gareth Stevens Publishing**
2544 Clinton Street
Buffalo, NY 14224



Designer: Andrea Davison-Bartolotta
Editor: Kristen Nelson

Photo credits: Cover, p. 1 (background) Moab Republic/Shutterstock.com; cover (dog) Nynke van Holten/shutterstock.com; p. 3 (dog) Wasitt Hemwarapornchai/Shutterstock.com; p. 3 (gerbil) Jearu/Shutterstock.com; p. 3 (raccoon) Saptashaw Chakraborty/Shutterstock.com; p. 3 (tiger) Astrid Lenz/Shutterstock.com; p. 5 Julia Zavalishina/Shutterstock.com; p. 7 Witsawat.S/Shutterstock.com; p. 9 WildLens Photograph/Shutterstock.com; p. 11 Laksika555/Shutterstock.com; p. 13 Petr Masek/Shutterstock.com; p. 15 Vladimir Vozdvizhenskiy/Shutterstock.com; p. 17 Ondrej Prosicky/Shutterstock.com; p. 19 MiViK/Shutterstock.com; p. 21 Svetlana Chernyshova/Shutterstock.com; p. 23 (bird) clarst5/Shutterstock.com; p. 23 (cat) Irina Gutyryak/Shutterstock.com; p. 23 (frog) Hintau Aliaksei/Shutterstock.com; p. 23 (lizard) PetlinDmitry/Shutterstock.com.



Printed in China

CPSIA compliance information: Batch #QSGS26: For further information contact Gareth Stevens at 1-800-542-2595.